Generis

PUBLISHING

Applications of Control Error Convergence Proof Using Lyapunov Direct Method Approach

for a Class of Mixed Fractional Order Model Reference Adaptive Control

Gustavo E. Ceballos
Manuel A. Duarte

Title: Applications of Control Error Convergence Proof Using Lyapunov Direct Method Approach

for a Class of Mixed Fractional Order Model Reference Adaptive Control

ISBN: 979-8-89966-458-8

Author: Gustavo E. Ceballos, Manuel A. Duarte

Cover image: www.pixabay.com

Publisher: Generis Publishing
Online orders: www.generis-publishing.com
Contact email: info@generis-publishing.com

Applications of Control Error Convergence Proof Using Lyapunov Direct Method Approach for a Class of Mixed Fractional Order Model Reference Adaptive Control

Gustavo E. Ceballos Benavides, Manuel A. Duarte-Mermoud

Abstract: This article, without being extensive in any way, compiles some important theoretical advances and application examples, applied to fractional control but mainly, it is an extension of one of our papers published in 2025, in which the Lyapunov stability theory was extended to mixed fractional-order direct model reference adaptive control (FO-DMRAC), where the adaptive control parameter is of fractional order, and the control error model is of integer order. In addition, examples are shown that help to consolidate this theorical result. We have primarily focused on the use of direct adaptive control due to its ease of implementation. However, other methods are not necessarily excluded, as long as they maintain a structure similar to the case where the plant is of integer order, and the adjustment laws are of fractional order. Finally, we present illustrative examples that we hope will allow the reader to gain more confidence in using these theoretical tools, developed over time, in the practical application of these fractional-order adaptive algorithms.

Prof. Dr. Gustavo E. Ceballos Benavides
Facultad de Ingeniería y Arquitectura, Universidad Central de Chile, Av. Santa Isabel 1186, Santiago 8330601, Chile.
e-mail: gustavo.ceballose@ucentral.cl

Prof. Dr. Manuel A. Duarte-Mermoud
Facultad de Ingeniería y Arquitectura, Universidad Central de Chile, Av. Santa Isabel 1186, Santiago 8330601, Chile.
e-mail: manuel.duarte@ucentral.cl

Acknowledgments

First and foremost, I want to thank my wife Andrea for her patience while I wrote this book. I also want to thank my children Natalia, Gustavo Jr., Arantzazu, Antonia, and Magdalena for their constant support. And of course, my sister Marcela for her unwavering support. As for the list of friends and colleagues who have supported me, it would be endless, which is why I will probably leave many out. However, I do want to highlight the support of my co-author Manuel Duarte and my friend Alfonso Ehijo for their constant assistance during the writing of this book.

Gustavo E. Ceballos

CONTENTS

1. Introduction

One of the objectives of this article is to show that the control error $e_1(t)$ for some class of fractional order dynamic systems such as the fractional order direct model reference adaptive control (FO-DMRAC) tend to zero as time tends to infinite using the classical Lyapunov stability approach.

The idea is not to show the performance of the controllers, but rather to show that although some of the conditions of the Lyapunov method ($\dot{V}(e,\phi) \leq 0$) are not met for all t, the control error $e_1(t) = y_p(t) - y_m(t)$ converges to 0 as long as the limit of $\dot{V}(e,\phi)$ tends to zero as t tends to infinity [1], where $y_p(t)$ is the plant output and $y_m(t)$ is the reference model output.

It is important to mention that the classic Lyapunov method was used, avoiding conceptual complexities as much as possible.

Many of the approaches applied to fractional adaptive systems use fractional (and not integer) derivatives of the Lyapunov function, achieving at most, demonstrate stability of the control error, but not convergence of this error to 0 [2-8].

Furthermore, many of these approaches are applied to fractional order systems with adaptive laws also of fractional order, facilitating the stability analysis for such systems. However, in most cases, the systems (or the control error dynamics) are of integer order and not fractional. Using the aforementioned approaches, it has only been possible to establish that the limit of the integral of the mean square error to infinite grows at most with a certain speed ($\lim_{t\to\infty}\left[t^{\alpha-\varepsilon}\frac{\int_{t_0}^{t} e(\tau)^2 d\tau}{t}\right] = 0,\ \forall\varepsilon > 0$), which is useful but does not allow to establish convergence of the control error itself to 0 [5,6].

Other approaches have used the gradient technique [9-17], but the error model dynamic is of fractional order [9-11].

Finally, we have used the adaptive error model 3 in which there is no access to the system's states, making the analysis more complex. Moreover, we have only considered model of unity relative degree ($n^* = 1$) and whose zeros belong to the rigth half complex s plane.

The organization of the article is as follows. Section 2 presents the classical or integer order implementation of the direct model reference adaptive control (IO-DMRAC). Section 3 shows some basic concepts and important theoretical results of the fractional calculus which will be the basis for analyzing the FO-DMRAC systems stability. In Section 4 various examples are presented that show the convergence to 0 of the control error and the behavior of the $\dot{V}(e,\phi)$ function. Finally, in Section 5 some conclusions are drawn.

2. Control Model System

In this section, we present the basic, well-known structure of the classical, or integer-order direct model reference adaptive control (IO-DMRAC) or simply DMRAC. Figure 2 shows a block diagram for the classical or integer-order Direct Model Reference Adaptive Control, in which control parameters are adjusted using their corresponding adaptive laws to keep the control error as small as possible.

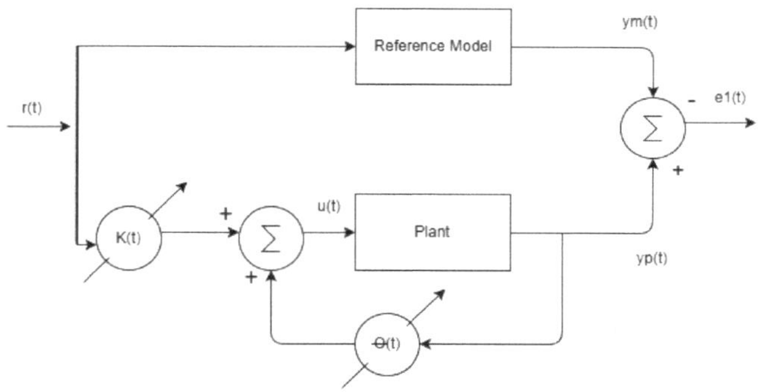

Figure 2. Classical DMRAC Block diagram.

2.1. DMRAC Algorithm

The objective of the MRAC is to minimize the control error. In the DMRAC approach, controller parameter adjustment is not required to ensure asymptotic convergence to the ideal parameters, making implementation more straightforward because the identification block is avoided [18].

Therefore, the DMRAC adaptive control scheme of a linear (or linearized) nth order plant, with relative degree 1, is shown in Figure 4, where only the control error is accessible (Error Model 3) but not the whole state error vector (Error Model 2), in which case, the analysis for determining stability conditions is simpler [18].

Figure 3 shows the block diagram implementation of the classical DMRAC in more detail.

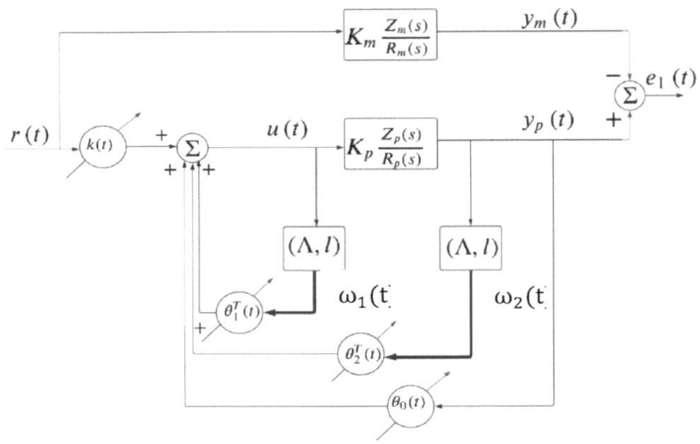

Figure 3. DMRAC block diagram for Plant relative degree is 1 ($n^* = 1$).

where $W_m(s) = k_m \frac{Z_m(s)}{R_m(s)}$ and $W_p(s) = k_p \frac{Z_p(s)}{R_p(s)}$ are the model reference and plant transfer function respectively.

Besides, the control law has the form

$u(t) = \theta^T(t) \cdot \omega(t)$ where $\theta(t) = [k(t), \theta_1^T, \theta_0(t), \theta_2^T]^T \in \mathbb{R}^{2n}$ and

$\omega(t) = \left[r(t), \omega_1^T, y_p(t), \omega_2^T\right]^T \in \mathbb{R}^{2n}$ are the controller parameters and the auxiliary signals respectively, and n, is the order of the plant.

$$\phi(t) = \begin{bmatrix} \psi(t) \\ \phi_1(t) \\ \phi_0(t) \\ \phi_2(t) \end{bmatrix} = \begin{bmatrix} k(t) - k^* \\ \theta_1(t) - \theta_1^* \\ \theta_0(t) - \theta_0^* \\ \theta_2(t) - \theta_2^* \end{bmatrix} \in \mathbb{R}^{2n}$$ are the parameters error vector controller

$\theta^* = \begin{bmatrix} k^* \\ \theta_1^* \\ \theta_0^* \\ \theta_2^* \end{bmatrix}$ are the ideal controller parameters.

The auxiliary signals are defined by

$$\dot{\omega}_1(t) = \Lambda\omega_1 + \ell u(t),$$

$$\dot{\omega}_2(t) = \Lambda\omega_2 + \ell y(t),$$

with $k(t), \theta_0(t), r(t), y_p(t) \in \mathbb{R}$, $\theta_1(t), \theta_2(t), \omega_1(t), \omega_2(t) \in \mathbb{R}^{n-1}$ and (Λ, ℓ) is any arbitrary stable and controllable pair, with $\Lambda \in \mathbb{R}^{n-1xn-1}$ a Hurwitz matrix.

For simplicity, we choose (Λ, ℓ) in the controllable canonical form. Furthermore, when $n^* = 1$, the control parameters for the classical or integer order adaptive laws (IO-DMRAC) can be chosen as:

$\dot{k}(t) = -sgn(k_p)e_1(t)r(t),$

$\dot{\theta}_0(t) = -sgn(k_p)e_1(t)y_p(t),$

$\dot{\theta}_1(t) = -sgn(k_p)e_1(t)\omega_1(t),$

$\dot{\theta}_2(t) = -sgn(k_p)e_1(t)\omega_2(t).$

and the output control error $e_1(t) = y_p(t) - y_m(t)\ \mathbb{R}$, can be expressed as $e_1(t) = \frac{k_p}{k_m} W_m(s)\phi^T(t)\omega(t)$, where $W_m(s)$ is a strictly positive real (SPR) transfer function.

On the other hand, for the fractional order adaptive laws (FO-DMRAC) case, these adaptive laws can be written as

$$ {}_{t_0}^{C}D_t^{\alpha}\theta(t) = {}_{t_0}^{C}D_t^{\alpha}\phi(t) = -sgn(k_p)e_1(t)\omega(t) \quad (1)$$

Finally, to obtain the values of the adaptive controller parameters analytically, we use the next relations.

$$k^* = \frac{k_m}{k_p}, \qquad \theta_1^{*T}(sI - \Lambda)^{-1}\ell = \frac{Z_m(s) - Z_p(s)}{Z_m(s)}, \quad \theta_0^* + \theta_2^{*T}(sI - \Lambda)^{-1}\ell = \frac{R_p(s) - R_m(s)}{k_p Z_m(s)}.$$

3. Fractional Calculus Preliminaries

In this section we present some definitions and the main advances in the stability of fractional order model reference adaptive control systems.

3.1 Basic concepts of fractional calculus

The basic definitions of fractional derivative and integral most used in engineering are presented [19,20] that will be useful for the implementation of the FO-DMRAC control.

Definition 1 [20]: The Riemann-Liouville fractional integral of order $\alpha > 0$ of a function $f(t) \in \mathbb{R}$ is defined by

$$I_{t_0}^{\alpha} f(t) = \frac{1}{\Gamma(\alpha)} \int_{t_0}^{t} \frac{f(\tau)}{(t-\tau)^{1-\alpha}} d\tau \text{ , } t > t_0 \text{ and } \mathcal{R}e(\alpha) > 0, \tag{2}$$

where $\Gamma(\alpha)$ is the Gamma function defined as

$$\Gamma(\alpha) = \int_0^{\infty} t^{\alpha-1} e^{-t} dt \text{ .} \tag{3}$$

Definition 2 [20]: Let $\alpha \geq 0$ and $[n] = \alpha$. The Caputo fractional derivative of order α of a function $f(t) \in \mathbb{R}$ is defined as

$$ {}_{t_0}^{C}D_t^{\alpha} f(t) = \frac{1}{\Gamma(n-\alpha)} \int_{t_0}^{t} \frac{f^{(n)}(\tau)}{(t-\tau)^{\alpha-n+1}} d\tau; \text{ as long as } f^{(n)} \in L_1[t_0, t]. \tag{4}$$

Some additional lemmas and a theorem are important for the stability analysis of fractional order adaptive control systems, which will be mentioned and whose proofs will be referenced in what follows.

Lemma 1: (Principle of fractional comparison): Let $e(t) \in \mathbb{R}^n$ be a vector of differentiable functions. Then, $\forall\ t \geq t_0$ the following inequality holds [2,3],

${}_{t_0}^{C}D_t^{\alpha}\{e^T(t)Pe(t)\} \leq 2e^T(t)P\,{}_{t_0}^{C}D_t^{\alpha}e(t), \forall\alpha \in (0,1]$, where $P \in \mathbb{R}^{nxn}$ is a symmetric square matrix of constant coefficients and positive definite. Proof of this Lemma can be found in [4].

3.2 Principal advances in stability of fractional order systems

In this subsection, we present several important theoretical results useful for fractional adaptive control.

Theorem 1: Let the state error $e(t)$ and the control error $e_1(t)$ be represented by equations

$$ {}_{t_0}^{C}D_t^{\beta}e(t) = A_{mn}e(t) + b_{mn}[\phi^T(t)\omega(t)], \qquad e(t_0) = e_0, \tag{5}$$

$$ e_1(t) = h_{mn}^T e(t), \qquad e_1(t_0) = e_{10}, $$

where $A_{mn} \in \mathbb{R}^{nxn}$ is a Hurwitz matrix and such that given a matrix $Q = Q^T > 0 \in \mathbb{R}^{nxn}$.

Then, there exists a matrix $P = P^T > 0 \in \mathbb{R}^{nxn}$ such that

$$ A_{mn}^T P + PA_{mn} = -Q, $$

$$ Pb_{mn} = h_{mn} $$

whose adaptive adjustment laws, to estimate the unknown controller parameters, are given by

$$ {}_{t_0}^{C}D_t^{\alpha}\phi(t) = {}_{t_0}^{C}D_t^{\alpha}\theta(t) = -\gamma sgn(k_p)e_1(t)\omega(t), \qquad \phi(t_0) = \phi_0 \tag{6}$$

where $k_p > 0$ is the gain of the plant, which is unknow, but the sign is known. Also, $\alpha < \beta$ and $\alpha \in (0,1]$. Then, if $e(t)$ and $\phi(t)$ are differentiable and uniformly continuous functions, it holds that

a) The parametric error $\phi(t)$, the state error $e(t)$ and the control error $e_1(t)$ remain bounded for all time.
b) Furthermore, if the auxiliary signal $\omega(t)$ is bounded, then ${}_{t_0}^{C}D_t^{\alpha}\phi(t)$ and ${}_{t_0}^{C}D_t^{\beta}e(t)$ also remain bounded.
c) The mean value of the squared norm of the state error $\overline{\|e(t)\|^2}$ is $o(t^{\varepsilon-\alpha})$ $\forall \varepsilon > 0$, or equivalently

$$\lim_{t\to\infty}\left[t^{\alpha-\varepsilon}\frac{\int_{t_0}^{t} e(\tau)^2 d\tau}{t}\right] = 0, \ \forall \varepsilon > 0 \tag{7}$$

where $o(t^{\varepsilon-\alpha})$ means that the speed of converges to zero is higher than $t^{-\alpha}$. The proof of this theorem can be found in [4].

Remark 1: This theorem can be applied to system whose relative degree n^* is grater than 1 as long as the model transfer function $W_m(s)$ is strictly positive real. Otherwise, it is necessary to modify $W_m(s)$ so that it meets this condition.

From Theorem 1, since $e_1(t) = h_{mn}^T e(t)$ with h_{mn}^T a constant vector, then, the control error $e_1(t)$ also will be $(t^{\varepsilon-\alpha})$ $\forall \varepsilon > 0$. Moreover, if (c) holds, it must also hold for the mean value of the square norm of $e_1(t)$, since $e_1(t) = h_{mn}^T e(t)$ with h_{mn}^T a vector whose components are constants.

There is a lemma that relaxes the hypothesis (b) imposed from Theorem 1 when $\beta = 1$ (i.e. the error model equation is of integer order), therefore all the internal signals $\omega(t)$ are bounded and there is no need to impose the boundedness condition over $\omega(t)$. The proof of this lemma can be found in [21]. Furthermore, as the auxiliary signal $\omega(t)$ is bounded and Theorem 1 guarantees (c), then the squared norm of the control error $|e_1(t)|^2$, also tends to 0 as t tends to infinity. That is

$$\lim_{t\to\infty}\left[t^{\alpha-\varepsilon}\frac{\int_{t_0}^{t} e_1(\tau)^2 d\tau}{t}\right] = 0,\ \ \forall\varepsilon > 0 \tag{8}$$

Therefore, the stability of the proposed FO-DMRAC is guaranteed.

Nevertheless, it is not possible to conclude convergence of the errors ($e(t)$ and $e_1(t)$) to 0 as t tends to ∞. Also, it is still a pending issue to prove the analytical differentiability of $\phi(t)$.

The high frequency gain of the plant b_p is supposed to be unknown but its sign is assumed to be known ($sgn(b_p) > 0$).

In the very well-known classical (or integer order) DMRAC, the proof of the convergence of the state error and the control error ($e(t)$ and $e_1(t)$ respectively) to 0 rests on the Barbalat Lemma [18]. That is, if the derivative of the Lyapunov function

$\dot{V}(e,\phi) = -e^T(t)Qe(t) \leq 0.$

Then, using Barbalat Lemma we can conclude that the state error tends to 0 ($\lim_{t\to\infty} e(t) = 0$) and therefore, the control error $e_1(t) = h_{mn}^T e(t)$ also tend to 0 ($\lim_{t\to\infty} e_1(t) = 0$). Nevertheless, even if it is not explicitly mentioned in the literature, because of the above, it must also be satisfied that $\lim_{t\to\infty} \dot{V}(e,\phi) = 0$.

In the case of mixed fractional order direct model reference adaptive control (FO-MRAC), there is not an equivalent Barbalat Lemma, therefore, it is not possible to conclude the convergence of the state error $e(t)$ and the control error $e_1(t)$ to 0.

Next, to aid the reader's understanding, we will present the approach for demonstrating convergence of the control error to 0 in the case of fractional-order adaptive control.

If we write the Lyapunov function for the mixed adaptive system, we have

$$V(e,\phi) = e^T(t)Pe(t) + \phi^T(t)\phi(t) \tag{9}$$

therefore,

$$\dot{V}(e,\phi) = -e^T Qe + 2\phi^T(t)\{e_1(t)\omega(t) - D^{1-\alpha}(e_1(t)\omega(t))\} \tag{10}$$

As

we can see, it is not easy to know the sign of the second term

$2\phi^T(t)\{e_1(t)\omega(t) - D^{1-\alpha}(e_1(t)\omega(t))\}$ or the term

$$\phi^T(t)\{e_1(t)\omega(t) - D^{1-\alpha}(e_1(t)\omega(t))\} \tag{11}$$

with $\alpha\epsilon(0,1)$. To make the notation easier, we can make the variable change

$\gamma = 1 - \alpha$, therefore $\gamma\epsilon(0,1)$. Then, Equation 11 could be re-written as

$$\phi^T(t)\{e_1(t)\omega(t) - D^{\gamma}(e_1(t)\omega(t))\} \tag{12}$$

By

Lemma 2 [11,12] we know that $\omega(t)$ is boundness. By part (a) of Theorem 1, $e_1(t)$ is boundness, therefore, $D^{\gamma}(e_1(t)\omega(t)) = 0$ (when t tends to infinity) by Caputo derivative property. Moreover, as $\phi(t)$ is boundness by the same theorem. Then, the Equation 12 is a boundedness term. Still, we cannot prove that the term

$$\dot{V}(e,\phi) = -e^T Qe + 2\phi^T(t)\{e_1(t)\omega(t) - D^{\gamma}(e_1(t)\omega(t))\} \leq 0$$

Remark 2: If the signals are of the PE (Persisting Excitation) class of an adequate order, $\lim_{t\to\infty} \phi(t)$ could converge to 0, but this is more related to the case of identification problems [18,22]. In the direct model reference adaptive control (DMRAC), not necessarily $\lim_{t\to\infty} \phi(t) = 0$.

Even in this case, we can only show that $\lim_{t\to\infty} \dot{V}(e,\phi) = \lim_{t\to\infty} -e^T(t)Qe(t)$ but it is not enough to conclude that $\lim_{t\to\infty} \dot{V}(e,\phi) = 0$ or $\lim_{t\to\infty} -e^T(t)Qe(t) = 0$ and thus, also conclude that $\lim_{t\to\infty} e(t) = 0$. Remember that Q is a constant and symmetric matrix.

Therefore, we need to explore what happens with the term inside the parenthesis. That is, the term $\{e_1(t)\omega(t) - D^\gamma(e_1(t)\omega(t))\}$.

First, we know that all the signals inside the parenthesis are bounded, then

$\lim_{t\to\infty}\{e_1(t)\omega(t) - D^\gamma(e_1(t)\omega(t))\} = \lim_{t\to\infty} e_1(t)\omega(t)$ because as we show before, by Caputo derivative property [19,20], $D^\gamma(e_1(t)\omega(t)) = 0$, as t tends to ∞, because the term $e_1(t)\omega(t)$ will be at least constant. Then, as $\omega(t)$ is a bounded signal, if $\lim_{t\to\infty} e_1(t) = 0$ then $\lim_{t\to\infty} \dot{V}(e,\phi) = 0$ and also $\lim_{t\to\infty} -e^T Qe = 0$ and therefore $\lim_{t\to\infty} e = 0$ and that will be the prove for the convergence to 0 of all errors (state error $e(t)$ and control error $e_1(t)$). But we are assumed that $\lim_{t\to\infty} e_1(t) = 0$ which is not possible to establish a priori.

It is important to mention that several fractional adaptive systems present this behavior. That is, $\lim_{t\to\infty} \dot{V}(e,\phi) = 0$ and $\lim_{t\to\infty} e_1(t) = 0$.

In the simulation results section (Section 4) we will show some examples that show this behavior for both, stable and unstable systems.

Now, from Equation (7), as ε is arbitrary, as long as $\varepsilon > 0$ we can choose $\varepsilon = \alpha$, therefore, Equation (7) can be re-written as

$\lim_{t\to\infty}\left[\frac{\int_{t_0}^{t} e(\tau)^2 d\tau}{t}\right] = 0$, which give a term that is divide by ∞. Even more, as part (a) of Theorem 1, $e(t)$ is boundedness or by part (b) of the same theorem, $\dot{e}(t)$ is boundedness and as $e(t)$ is part of the Equation 7, where all the other terms are boundedness, then $e(t)$ must be boundedness, therefore the numerator of Equation 7 or the term $\lim_{t\to\infty}\int_{t_0}^{t} e(\tau)^2 d\tau$ can be infinite or constant (if $e(t)$ is square integrable), then, we can use L'Hopital rule, that is

$$\lim_{t\to\infty}\left[\frac{\int_{t_0}^{t} e(\tau)^2 d\tau}{t}\right] = \lim_{t\to\infty}\left[\frac{\frac{d}{dt}\left(\int_{t_0}^{t} e(\tau)^2 d\tau\right)}{\frac{d}{dt}(t)}\right] = \lim_{t\to\infty}\left[\frac{e(t)^2}{1}\right] = \lim_{t\to\infty} e(t)^2 = 0 \Rightarrow \lim_{t\to\infty} e(t) = 0.$$

Then, since $e_1(t) = h_{mn}^T e(t)$ with h_{mn} a constant vector, then $\lim_{t\to\infty} e_1(t) = 0$ which mean that the control error $e_1(t)$ tends to 0 conform t tends to ∞. Therefore, we have proven the convergence of the state and control error to 0 which is new for fractional order direct model reference adaptive control systems (FO-DMRAC). Also, this result means that $\lim_{t\to\infty} \dot{V}(e,\phi) = 0$ as in the integer order case.

4. Simulation Results

In this section, we will show some examples that support what has been developed and conclude in Section 4. The simulations were performed using Matlab-Simulink [23-25]. For all examples, zero initial conditions were considered.

For simplicity of the analysis, but without loss of generality, $b_p = 1$ is considered. Then, a typical Lyapunov function can be choose as $V(e,\phi) = e^T(t)Pe(t) + \phi^T(t)\phi(t)$ therefore,

$\dot{V}(e,\phi) = -e^T Qe + 2\phi^T(t)\{e_1(t)\omega(t) - D^{1-\alpha}(e_1(t)\omega(t))\}$. If $\gamma = 1-\alpha$, then, the above term will be rewritten as

$$\dot{V}(e,\phi) = -e(t)^T Qe(t) + 2\phi^T(t)\{e_1(t)\omega(t) - D^{\gamma}(e_1(t)\omega(t))\} \tag{13}$$

Unlike the classic case, in equation (13), a term with a fractional derivative appears. For doing the simulation analysis, the designer can choose any $\alpha, \gamma \in (0,1]$ such as $\gamma = 1-\alpha$. In this case, we will choose $\alpha = 0.8 \Rightarrow \gamma = 0.2$.

In the mixed FO-DMRAC control case where the error model dynamics are of integer order, and the adaptive laws are fractional (Equation 1), the dynamics of the state error, including the plant and controller (error model 3) [18], is

$$\begin{gathered} \dot{e}(t) = A_{mn}e(t) + b_{mn}[\phi^T(t)\omega(t)] \\ e_1(t) = h_{mn}^T e(t) \end{gathered} \tag{14}$$

where

$$A_{mn} = \begin{bmatrix} A_p + b_p\theta_0^* h_p^T & b_p\theta_1^{*T} & b_p\theta_2^{*T} \\ \ell\theta_0^* h_p^T & \Lambda + l\ell & \ell\theta_2^{*T} \\ \ell h_p^T & 0 & \Lambda \end{bmatrix} \quad b_{mn} = \begin{bmatrix} b_p \\ \ell \\ 0 \end{bmatrix}$$

and

$h_{mn} = \begin{bmatrix} h_p^T & 0 & 0 \end{bmatrix}^T$, $x(t) = \begin{bmatrix} x_p^T(t) & \omega_1^T(t) & \omega_2^T(t) \end{bmatrix}^T$ and

$x_{mn}(t) = \begin{bmatrix} x_p^{*T}(t) & \omega_1^{*T}(t) & \omega_2^{*T}(t) \end{bmatrix}^T$, where $e(t) = x(t) - x_{mn}(t)$.

Although, from a theoretical point of view, it is not necessary to calculate the value of P from Q since it is enough that the conditions of the Meyer-Kalman-Yakubovich (MKY) lemma are met, it is important for the purposes of carrying out the calculation of $\dot{V}(e,\phi)$.

Next, we present some examples to show the behavior of the control error $e_1(t)$ and the derivative $\dot{V}(e,\phi)$ of the Lyapunov function that confirm the convergence to 0 of $e_1(t)$ which is the main result of this article, and which was proven in the previous section. In all the examples shown below, the fractional parameters α_i of Equation 6, has been choose as $\alpha_i = 0.8$ for simplicity of analysis. The designer can choose the α_i values as desired or through an optimization process that could ensure a better response of the plant under control. Nevertheless, we need not to forget that $\alpha_i \in (0,1)$. Also, the values of the adaptive gains $\Gamma_i > 0$ have been fixed to 1, but, newly, the designer can choose the values as desired.

4.1. Scalar stable plant

Consider that n is the order of the plant and n^* is its relative degree.

Example 1: ($n = 1$ and $n^* = 1$)

Table 1: FO-DMRAC implementation for a stable plant

Plant model	Reference Model
$\dot{x}_p(t) = -x_p(t) + u(t).$ $y_p(t) = x_p(t).$ $W_p(s) = \frac{1}{s+1}.$ $a_p = -1, b_p = 1, h_p = h_p^T = 1.$	$\dot{x}_m(t) = -2(t) + 2r(t).$ $r(t) = 1.$ $y_m(t) = x_m(t).$ $W_m(s) = \frac{2}{s+2}.$ $a_m = -2, b_m = 2, h_m = h_m^T = 1.$
$\Lambda = 0, \quad \ell = 0, \quad A_{mn} = -2,\ b_{mn} = -1,$ $e(t) = e_1(t) = x_p - x_m$	

Remark 3: In the scalar case, $\theta_0^* = \theta^* = \frac{a_m - a_p}{b_p}$ and $k^* = \frac{b_m}{b_p}$ where θ^* and k^* are the ideal control parameters and $u(t) = k(t)r(t) + \theta(t)x_p(t)$ is the control law.

Then, using MKY lemma, that is, for any $Q = Q^T > 0$ exist a $P = P^T > 0$ such that

$$A_{mn}^T P + P A_{mn} = -Q.$$

$$P b_{mn} = h_{mn}.$$

we may calculate A_{mn}, P, and Q for the system of Example 1 and thus, be able to calculate $\dot{V}(e, \phi)$.

In this example, $A_{mn} = -2, P = 1, Q = 2, k^* = \frac{b_m}{b_p} = 2$ and $\theta^* = \frac{a_m - a_p}{b_p} = -1.$

The fractional adaptive laws were implemented using the Ninteger Toolbox [24]. Specifically, the NID block was used based on the Oustaloup method [25].

Figure 4 shows the plant response $y_p(t)$, reference output $y_m(t)$ and control error $e_1(t)$ respectively. From Figure 5 we can see the evolution of $\dot{V}(e, \phi)$. respectively.

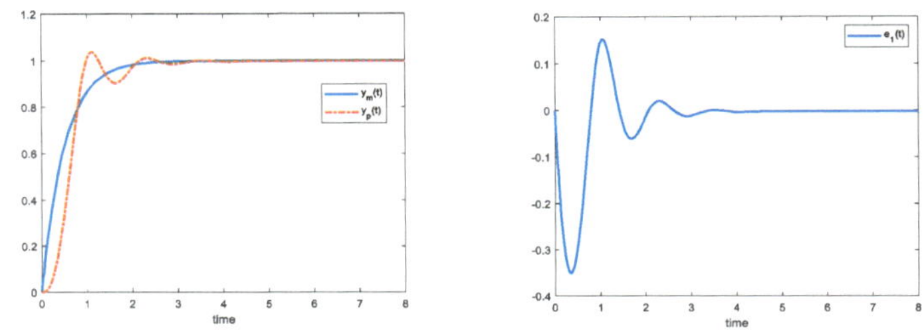

(a) (b)

Figure 4. (a) FO-DMRAC reference output, response and (b) control error when $\mathbf{r(t) = 1}$.

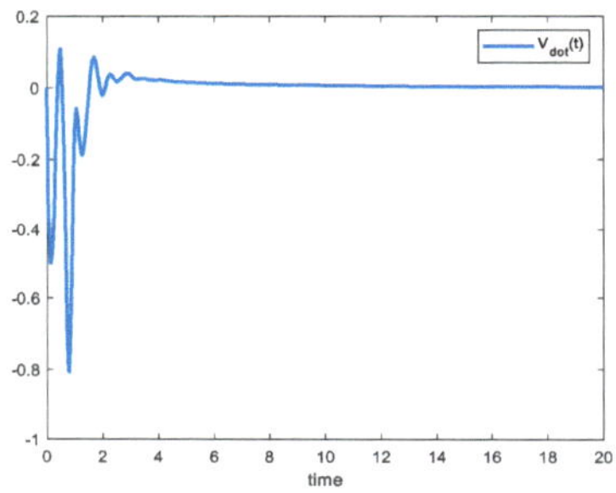

Figure 5. FO-DMRAC graph of $\dot{V}(e, \phi)$vs t when $\mathbf{r(t) = 1}$.

Also, from figure 4 and 5 it can be seen that the control error $e_1(t)$ converges to 0 and $\dot{V}(e, \phi)$ can be greater than 0 in some time intervals but also converge to 0 conform t tends to ∞ such as the classical case.

If we change $r(t)$ by a sinusoidal signal such as $r(t) = 2\sin(t)$, the response, control error and the graph of $\dot{V}(e, \phi)$ are shown in figures 6 and 7 respectively.

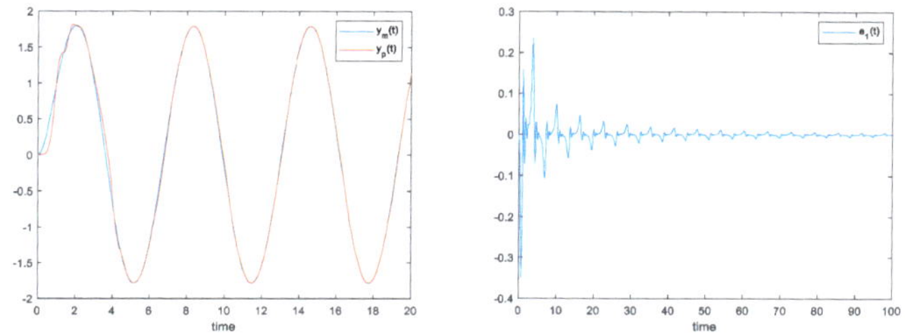

(a) (b)

Figure 6. (a) FO-DMRAC reference output, response and (b) control error when $\boldsymbol{r(t)} =$ $\boldsymbol{2sin\,(t)}$.

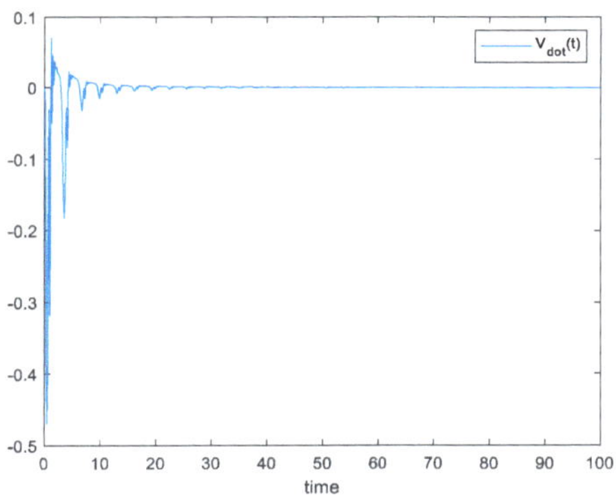

Figure 7. FO-DMRAC graph of $\boldsymbol{\dot{V}(e,\phi)}$ when $\boldsymbol{r(t) = 2sin\,(t)}$.

From Figure 6 it can be seen that the control error $e_1(t)$ convergence takes more time than the case when the reference is a step, but the convergence is assured anyway. Similar convergence behavior can be seen from Figure 7.

4.2. Scalar unstable plant

Example 2: ($n = 1$ and $n^* = 1$)

Table 2: FO-DMRAC implementation for an unstable plant

Plant model	Reference Model
$\dot{x}_p(t) = 1(t) + 1u(t).$ $y_p(t) = x_p(t).$ $W_p(s) = \frac{1}{s-1}.$ $a_p = 1, b_p = 1, h_p = h_p^T = 1.$	$\dot{x}_m(t) = -2(t) + 2r(t).$ $r(t) = 1.$ $y_m(t) = x_m(t).$ $W_m(s) = \frac{2}{s+2}.$ $a_m = -2, b_m = 2, h_m = h_m^T = 1.$
$\Lambda = 0, \quad \ell = 0, \quad A_{mn} = -2,\ b_{mn} = -1,$ $e(t) = e_1(t) = x_p - x_m$	

$A_{mn} = -2, P = 1,\ Q = 2,\ k^* = \frac{b_m}{b_p} = 2$ and $\theta^* = \frac{a_m - a_p}{b_p} = -3.$

The next figures show similar convergence behavior as the graphics of Example 1 for the unstable plant.

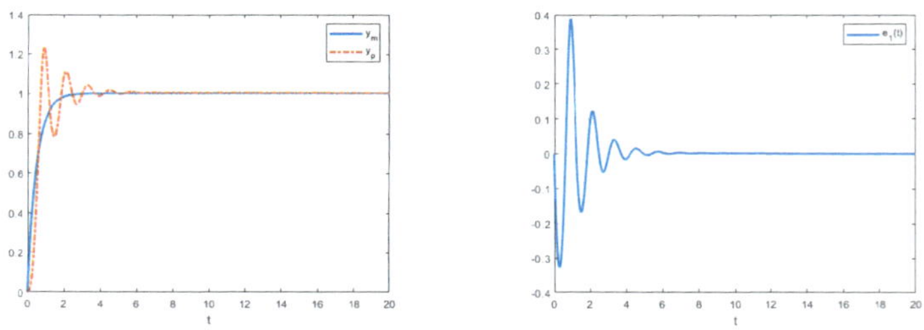

(a) (b)

Figure 8. (a) FO-DMRAC reference output, response and (b) control error when $r(t) = 1$.

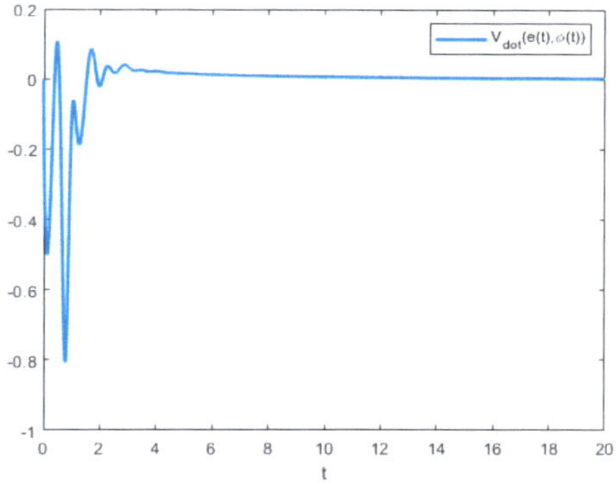

Figure 9. FO-DMRAC graph of $\dot{\mathbf{V}}(\boldsymbol{e},\boldsymbol{\phi})\boldsymbol{vs}\ \boldsymbol{t}$ ***when*** $\mathbf{r}(\boldsymbol{t}) = \mathbf{1}$.

Now, if we change the reference to $r(t) = 2\sin{(t)}$ the figures are the following (figures 10 and 11).

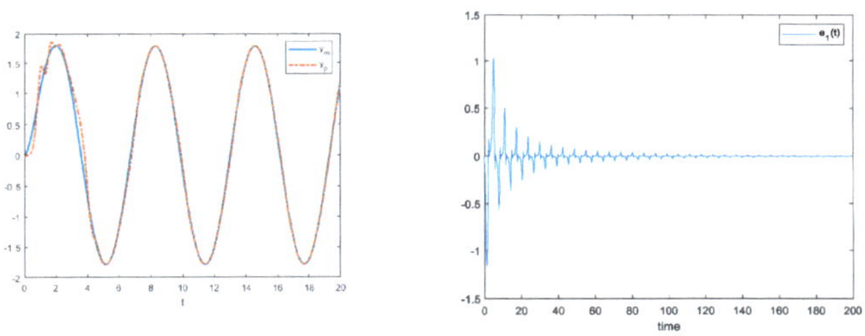

(a) (b)

Figure 10. (a) FO-DMRAC reference output, response and (b) control error when $\mathbf{r}(\boldsymbol{t}) = \mathbf{2}\boldsymbol{sin}\ (\boldsymbol{t})$.

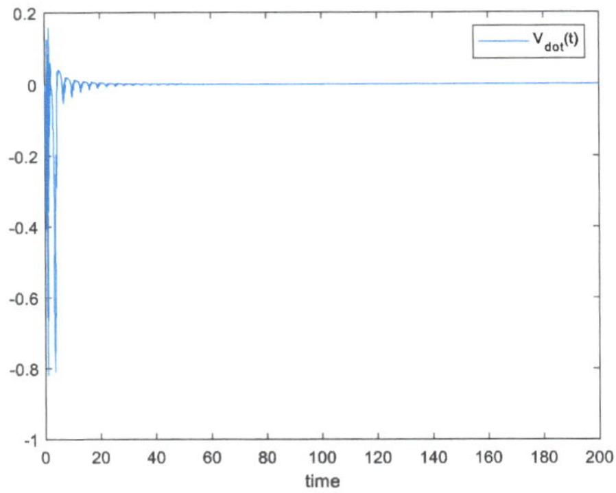

Figure 11. FO-DMRAC graph of $\dot{V}(e,\phi)$ when $r(t) = 2sin\ (t)$.

As we can see, from figures 10 and 11, both the control error $e_1(t)$ and $\dot{V}(e,\phi)$ converge to 0, even if $\dot{V}(e,\phi)$ can take positive values in some time intervals.

4.3. Vectorial case (stable plant)

Example 3: ($n = 2$ and $n^* = 1$)

Table 3: FO-DMRAC implementation

Plant model	Reference Model
$\dot{x}_p(t) = A_p(t)x_p(t) + b_p u(t).$ $y_p(t) = h_p^T x_p(t).$ $W_p(s) = \frac{s+1}{s^2+7s+12}$ Controllable canonical form: $A_p = \begin{bmatrix} 0 & 1 \\ -12 & -7 \end{bmatrix}, b_p = \begin{bmatrix} 0 \\ 1 \end{bmatrix},$ $h_p^T = [1\ 1]$	$\dot{x}_m(t) = -2x_m((t) + 2r(t).$ $y_m(t) = x_m(t).\ r(t) = 1.$ $W_m(s) = \frac{2}{s+2}\frac{(s+1)}{(s+1)}.$ Controllable canonical form: $A_m = \begin{bmatrix} 0 & 1 \\ -2 & -3 \end{bmatrix}, b_m = \begin{bmatrix} 0 \\ 1 \end{bmatrix},$ $h_{mn}^T = [h_p^T\ 0\ 0]$
$\Lambda = -1, \quad \ell = 1,\ b_{mn} = [0\ 1\ 1\ 0]^T, h_{mn}^T = [1\ 1\ 0\ 0]$ $e(t) = x(t) - x_{mn}(t),\ e_1(t) = h_{mn}^T e(t)$	

Remark 4: The reference model transfer function $W_m(s)$ has been changed (order 2) just to match the conditions for finding the control ideal parameters θ^*, but the behavior is the same as the original transfer function $W_m(s) = \frac{2}{s+2}$.

Control Laws:

$$\theta(t) = [k(t) \quad \theta_1^T(t) \quad \theta_0(t) \quad \theta_2^T(t)]^T \in \mathbb{R}^4$$

$$\omega(t) = \left[r(t) \quad \omega_1^T(t) \quad y_p(t) \quad \omega_2^T(t)\right]^T \in \mathbb{R}^4$$

$$u(t) = \theta^T(t)\omega(t)$$

Note: $\theta_1^T(t)$ and $\theta_2^T(t) \in \mathbb{R}$

Internal auxiliary signals:

$$\dot{\omega}_1(t) = \Lambda\omega_1(t) + \ell u(t)$$

$$\dot{\omega}_2(t) = \Lambda\omega_2(t) + \ell y_p(t)$$

$$\Lambda = -1$$

$$\ell = 1$$

Fractional order adaptive laws:

$${}_{t_0}^{C}D_t^{\alpha_i}\theta(t) = -\Gamma_i e_1(t)\omega(t)$$

Remark 5: Γ_i are the elements of the diagonal matrix Γ where $i = 1, \dots, 4$. For practical implementation simplicity, we choose $\Gamma = I$ (identity matrix of appropriated dimension).

$\phi(t) = \begin{bmatrix} \psi(t) \\ \phi_1(t) \\ \phi_0(t) \\ \phi_2(t) \end{bmatrix} = \begin{bmatrix} k(t) - k^* \\ \theta_1(t) - \theta_1^* \\ \theta_0(t) - \theta_0^* \\ \theta_2(t) - \theta_2^* \end{bmatrix} \in \mathbb{R}^4$ are the parameters error vector controller and

$\theta^* = \begin{bmatrix} k^* \\ \theta_1^* \\ \theta_0^* \\ \theta_2^* \end{bmatrix} = \begin{bmatrix} 2 \\ 0 \\ 4 \\ 6 \end{bmatrix}$ are the ideal controller parameters.

Finally, $A_{mn} = \begin{bmatrix} 0 & 1 & 0 & 0 \\ -8 & -3 & 0 & 6 \\ 4 & 4 & -1 & 6 \\ 1 & 1 & 0 & -1 \end{bmatrix}$

whose eigenvalues are -1, -2, -1 and -1 therefore A_{mn} is Hurwitz. Also, if

$Q = \begin{bmatrix} 2 & 26.5 & -23.5 & -6 \\ 26.5 & 33 & -16.5 & -8 \\ -23.5 & -16.5 & 4 & 0 \\ -6 & -8 & 0 & 2 \end{bmatrix}$ which is a symmetrical and positive defined matrix,

then

$P = \begin{bmatrix} 4 & 0.5 & 0.5 & 1 \\ 0.5 & 3 & -2 & 0 \\ 0.5 & -2 & 2 & 0 \\ 1 & 0 & 0 & 1 \end{bmatrix}$ is also a symmetrical and positive defined matrix that

satisfies the MKY Lemma. That is

$$A_{mn}^T P + P A_{mn} = -Q,$$

$$P b_{mn} = h_{mn}.$$

As before, figures 12 and 13 show the response, control error $e_1(t)$ and $\dot{V}(e, \phi)$ behavior respectively.

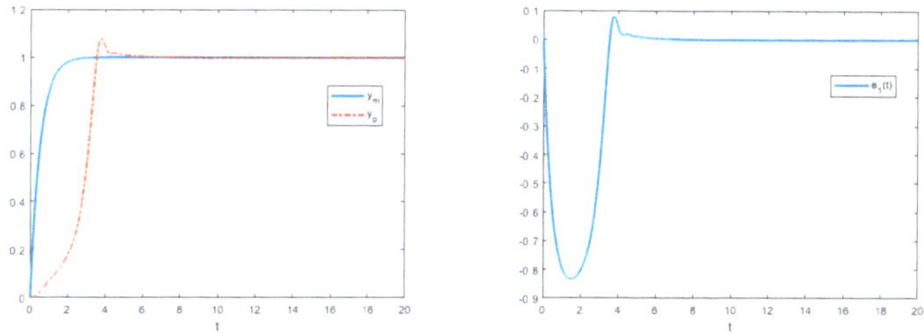

(a) (b)

Figure 12. (a) FO-DMRAC reference output, response and (b) control error when $\mathbf{r}(\boldsymbol{t}) = \mathbf{1}$.

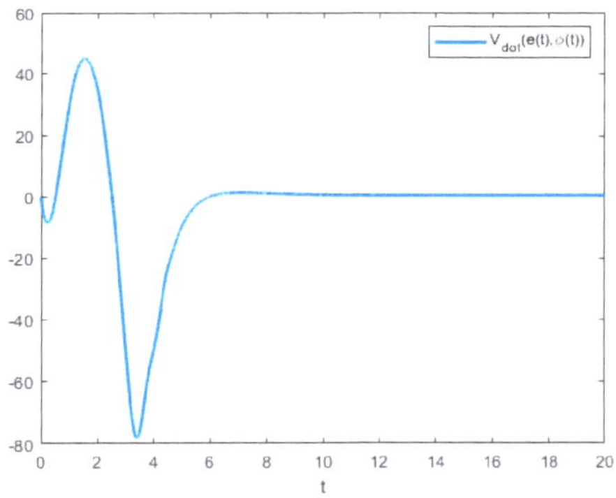

Figure 13. FO-DMRAC graph of $\dot{V}(\boldsymbol{e}, \boldsymbol{\phi})$ when $\mathbf{r}(\boldsymbol{t}) = \mathbf{1}$.

From Figure 12 we can see a similar behavior as examples 1 and 2, the plant output $y_p(t)$ converges to the model reference output $y_m(t)$ which is equivalent to saying that the control error tends to 0 as t tends to ∞. Also, from Figure 13 we confirm that $\lim_{t\to\infty} \dot{V}(e, \phi) = 0$ even if in some time intervals (mainly in the transient period), $\dot{V}(e, \phi)$ can be positive.

Now, by changing the reference signal to a sinusoidal one as in the previous examples, that is, $r(t) = 2\sin(t)$, we obtain the following dynamic behavior which are shown in the next figures.

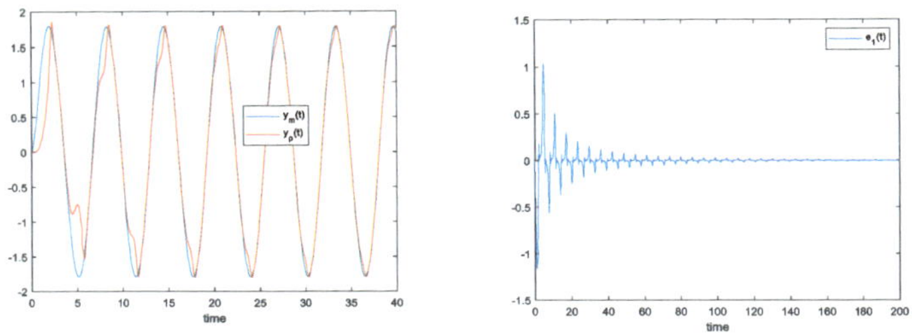

(a) (b)

Figure 12. (a) FO-DMRAC reference output, response and (b) control error when $\boldsymbol{r(t) = 2sin\,(t)}$.

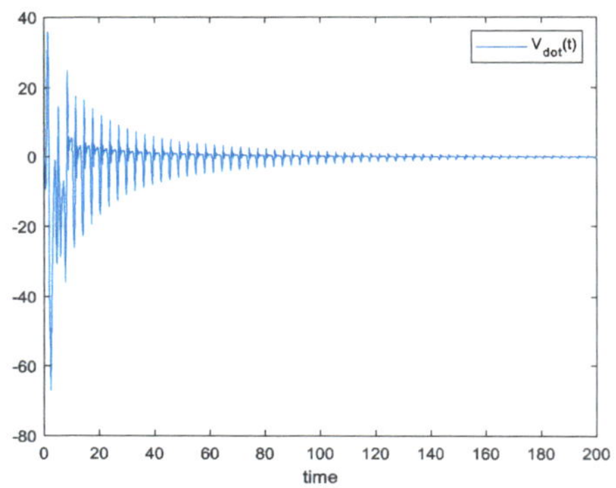

Figure 13. FO-DMRAC graph of $\boldsymbol{\dot{V}(e, \phi)}$ when $\boldsymbol{r(t) = 2sin\,(t)}$.

Remark 6: These simulations are made to show that $\lim_{t\to\infty} e_1(t) = 0$ (control error converge to 0) in the case of fractional order model reference adaptive control implementation (FO-DMRAC) and not to show the performance of the controllers. Better performance can be achieved by using an optimization algorithm such as PSO [26] or another one, in such a way as to choose the fractional parameters α_i and adaptive gains Γ_i more conveniently.

4.4. Vectorial case (unstable plant)

Example 4: ($n = 2$ and $n^* = 1$)

Table 3: FO-DMRAC implementation

Plant model	Reference Model
$\dot{x}_p(t) = A_p(t)x_p(t) + b_p u(t).$ $y_p(t) = h_p^T x_p(t).$ $W_p(s) = \frac{s+1}{s^2+s-12}$ Controllable canonical form: $A_p = \begin{bmatrix} 0 & 1 \\ 12 & -1 \end{bmatrix}, b_p = \begin{bmatrix} 0 \\ 1 \end{bmatrix},$ $h_p^T = [1\ 1]$	$\dot{x}_m(t) = -2x_m((t) + 2r(t).$ $y_m(t) = x_m(t).\ r(t) = 1.$ $W_m(s) = \frac{2}{s+2}\frac{(s+1)}{(s+1)}.$ Controllable canonical form: $A_m = \begin{bmatrix} 0 & 1 \\ -2 & -3 \end{bmatrix}, b_m = \begin{bmatrix} 0 \\ 1 \end{bmatrix},$ $h_{mn}^T = [h_p^T\ 0\ 0]$
$\Lambda = -1, \quad \ell = 1,\ b_{mn} = [0\ 1\ 1\ 0]^T, h_{mn}^T = [1\ 1\ 0\ 0]$ $e(t) = x(t) - x_{mn}(t),\ e_1(t) = h_{mn}^T e(t)$	

$$\theta^* = \begin{bmatrix} k^* \\ \theta_1^* \\ \theta_0^* \\ \theta_2^* \end{bmatrix} = \begin{bmatrix} 2 \\ 0 \\ -2 \\ -12 \end{bmatrix}$$ are the ideal controller parameters.

Finally, $$A_{mn} = \begin{bmatrix} 0 & 1 & 0 & 0 \\ 10 & -3 & 0 & -12 \\ -2 & -2 & -1 & -12 \\ 1 & 1 & 0 & -1 \end{bmatrix}$$

whose eigenvalues are -1, -2, -1 and -1 therefore A_{mn} is Hurwitz. Also, if

$$Q = \begin{bmatrix} -10 & -36.5 & 24.5 & 12 \\ -36.5 & 9 & -4.5 & 10 \\ 24.5 & -4.5 & 4 & 0 \\ 12 & 10 & 0 & 1 \end{bmatrix}$$ which is a symmetrical and positive defined matrix, then

$P = \begin{bmatrix} 4 & 0.5 & 0.5 & 1 \\ 0.5 & 3 & -2 & 0 \\ 0.5 & -2 & 2 & 0 \\ 1 & 0 & 0 & 1 \end{bmatrix}$ is also a symmetrical and positive defined matrix that

satisfies the MKY Lemma. That is

$$A_{mn}^T P + P A_{mn} = -Q,$$

$$P b_{mn} = h_{mn}.$$

As we can see from figures 14 and 15, the response, control error $e_1(t)$ and $\dot{V}(e, \phi)$ have a behavior that was predicted by the demonstration that both, $e_1(t)$ and $\dot{V}(e, \phi)$ tend to 0, even if $\dot{V}(e, \phi) > 0$ in some time periods.

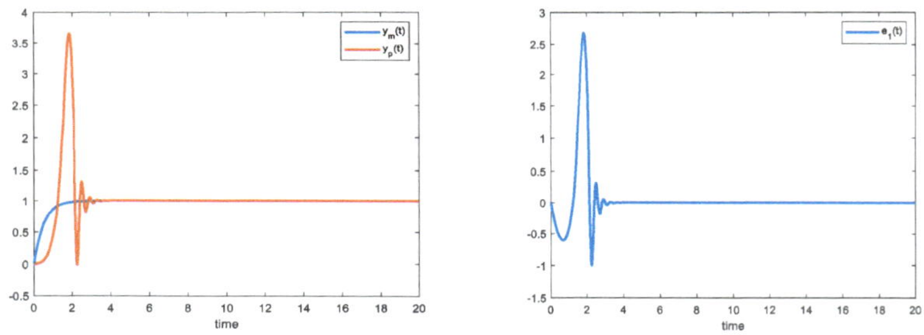

(a) (b)

Figure 14. (a) FO-DMRAC reference output, response and (b) control error when $\mathbf{r(t) = 1}$.

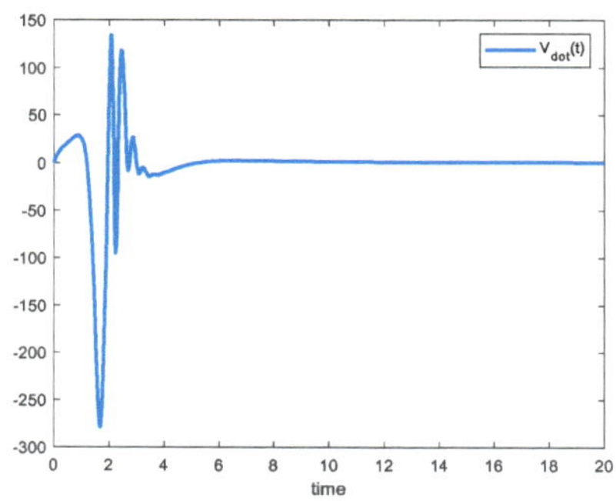

Figure 15. FO-DMRAC graph of $\dot{\mathbf{V}}(\mathbf{e}, \boldsymbol{\phi})$ ***when*** $\mathbf{r(t) = 1}$.

Nevertheless, in this example (unstable plant), the signals have more oscillations than in the stable case, newly, we can see the convergence of the control error $e_1(t)$ and the $\dot{V}(e,\phi)$ to zero.

Now, by changing the reference signal to a sinusoidal one as in the previous examples, that is, $r(t) = 2\sin(t)$, we obtain the following dynamic behavior which are shown in the next figures.

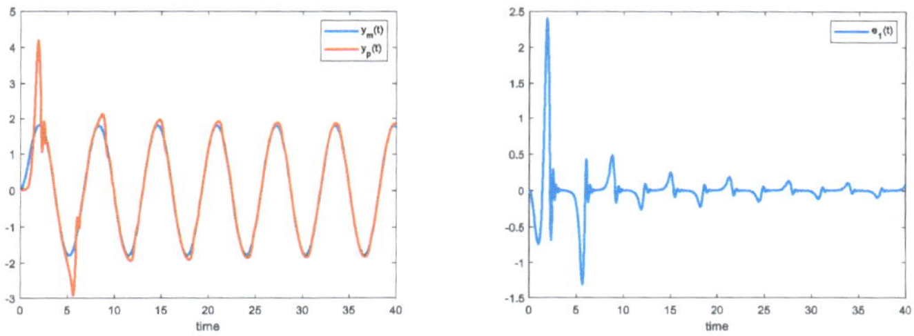

(a) (b)

Figure 16. (a) FO-DMRAC reference output, response and (b) control error when $\mathbf{r(t) = 2sin\ (t)}$.

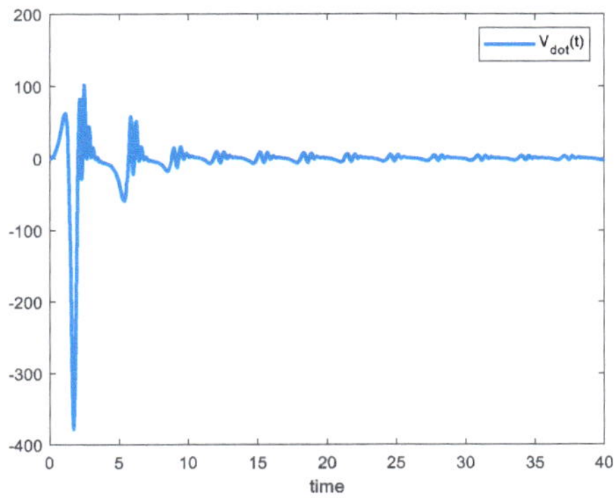

Figure 17. FO-DMRAC graph of $\dot{\mathbf{V}}(\boldsymbol{e},\boldsymbol{\phi})$ ***when*** $\mathbf{r(t) = 2sin\ (t)}$.

It is important to emphasize that the goal of this article is to show that the control error $e_1(t)$ converge to zero in the fractional case of the CARMD, without considering whether the plant output, when applying fractional adaptive control, is the best or closest to the reference output throughout all simulation time. To determine a better plant output, any optimization method (such as PSO or another heuristic method like Genetic Algorithms) can be used to determine the most suitable values for both, the adaptive gains and the orders of the fractional derivatives of the adaptive tuning laws.

Finally, new conditions of the Lyapunov method applied to FO-DMRAC can be established. That is, let be a $V(e,\phi)$ a continuous and derivable function, then, if

$$V(e,\phi) > 0$$

$$\dot{V}(0,0) = 0$$

$$\text{and if } \lim_{t\to\infty} \dot{V}(e,\phi) = 0,$$

then, $\lim_{t\to\infty} e(t) = 0$, which implies that the control error $e_1(t)$ also converges to 0. Therefore, we could conclude that the stability second method of Lyapunov is a special case of this new theoretical result, at least for DMRAC method.

Remark 7: It is interesting to note that an easy way to find the ideals controller parameters, that sometimes are difficult to find analytically, is to use a Persistent Excitation (PE) signal as reference to the controlled plant and wait until parametric convergence occurs. Of course, it is possible that a longer simulation time will be required to ensure the parametric convergence.

In the Figure 18, we show, in the case of the example 4 (unstable vectorial plant), as the application of a PE signal to the reference signal, can give us the ideal controller parameters.

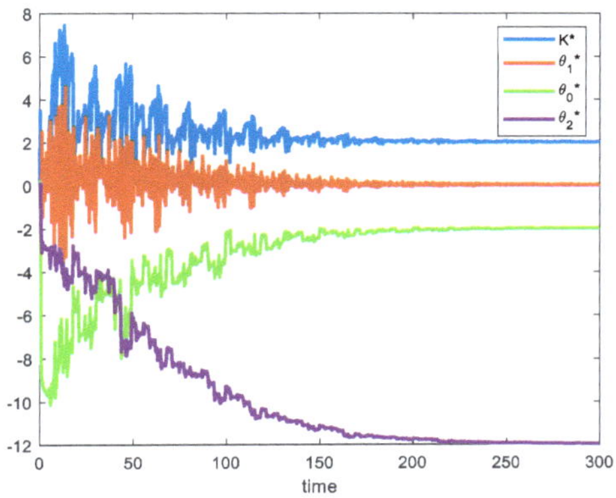

Figure 18: Determination of the parameters of the direct adaptive controller using a practical (non-analytical) method.

We can see, from the Figure 18, that the controller parameters are the same as the ones that were found analytically. That is to say,

$$\theta^* = \begin{bmatrix} k^* \\ \theta_1^* \\ \theta_0^* \\ \theta_2^* \end{bmatrix} = \begin{bmatrix} 2 \\ 0 \\ -2 \\ -12 \end{bmatrix}.$$

5. Conclusions

In this article, we have proven that in the FO-DMRAC case, the control error $e_1(t) = y_p(t) - y_m(t)$ converges to 0 even if the derivative of the Lyapunov function $V(e, \phi)$ is greater than 0 in some time intervals. Some examples have been presented to show the convergence of the control error $e_1(t)$ and the behavior of the function $\dot{V}(e, \phi)$ as t approaches infinity.

One of the main difficulties in carrying out the synthesis of $\dot{V}(e, \phi)$ for simulation purposes is that we need to compute the ideal control parameters θ^* analytically and the matrix Q explicitly, which is not an easy task when the order of the plant to control grows in order. One practical approach to determine the ideal controller parameters θ^* is to implement the controller system using a Persistent Excitation (PE) reference input signal and wait for convergence. Sometimes, it could take a long time depending on the order system because the higher the order, the more parameters we must compute.

These results (convergence of the control error $e_1(t)$ to zero) could be extended to plants with relative degree greater than unity as long as the model reference transfer function can be transformed is some way, to a strictly positive real transfer function. Moreover, plants whose zeros are located in the right half complex s plane are not yet approach in the fractional adaptive control literature.

In conclusion, the convergence of the state error $e(t)$ and the control error $e_1(t)$ to zero has been demonstrated for a class of fractional adaptive control systems (FO-DMRAC) whose relative degree is one and the zeros are located in the left half complex s plane. Previously, it had only been established that the mean square errors converge to zero as time approaches infinity; however, the convergence of control errors themselves had not been explicitly demonstrated until now [1].

References

[1] Ceballos Benavides, G.E.; Duarte-Mermoud, M.A.; Martell, L.B. Control Error Convergence Using Lyapunov Direct Method Approach for Mixed Fractional Order Model Reference Adaptive Control. Fractal Fract. **2025**, 9, 98. https://doi.org/10.3390/fractalfract9020098.

[2] Duarte-Mermoud, M.A.; Aguila-Camacho, N.; Gallegos, J.A.; Castro-Linares, R. Using General Quadratic Lyapunov Functions to Prove Lyapunov Uniform Stability for Fractional Order Systems. Commun. Nonlinear Sci. Numer. Simul. **2015**, 22, 650–659.

[3] Aguila-Camacho, N.; Gallegos, J.; Duarte-Mermoud, M.A. Analysis of fractional order error models in adaptive systems: Mixed order cases. Fract. Calc. Appl. Anal. **2019**, 22, 1113–1132. Fractal Fract. **2025**, 9, 98 16 of 16.

[3] Aguila-Camacho, N.; Duarte-Mermoud, M.A.; Gallegos, J.A. Lyapunov Functions for Fractional Order Systems. Commun. Nonlinear Sci. Numer. Simul. **2014**, 19, 2951–2957.

[5] Hu, J.-B.; Lu, G.-P.; Zhang, S.-B.; Zhao, L.-D. Lyapunov Stability Theorem about Fractional System Without and with Delay. Commun. Nonlinear Sci. Numer. Simul. **2015**, 2, 905–913.

[6] Balaska, H.; Ladaci, S.; Djouambi, A.; Schulte, H.; Bourouba, B. Fractional Order Tube Model Reference Adaptive Control for a Class of Fractional Order Linear Systems. Int. J. Appl. Math. Comput. Sci. **2020**, 30, 501–515.

[7] Hu, C.; Qi, Z.; Ma, Q.; Zhou, X. Factional Order Model Reference Adaptive Control based on Lyapunov Stability Theory. In Proceedings of the 35th Chinese Control Conference, Chengdu, China, 27–29 July **2016**.

[8] Zhang, R.; Liu, Y. A new Barbalat's Lemma and Lyapunov stability theorem for Fractional order Systems. In Proceedings of the 2017 29th Chinese Control and Decision Conference (CCDC), Chongqing, China, 28–30 May **2017**.

[9] Ladaci, S.; Charef, A. On fractional adaptive control. Nonlinear Dyn. **2006**, 43, 365–378.

[10] Petras, I. An Adaptive Fractional-Order Controller. In Proceedings of the 2013 14th International Carpathian Control Conference (ICCC), Rytro, Poland, 26–29 May **2013**; pp. 297–301.

[11] Vinagre, B.M.; Petras, I.; Podlubny, I.; Chen, Y. Using fractional order adjustment rules and fractional order reference models in model-reference adaptive control. Nonlinear Dyn. **2002**, 29, 269–279.

[12] Chen, Y.; Gao, Q.;Wei, Y.;Wang, Y. Study on fractional order gradient methods. Appl. Math. Comput. **2017**, 314, 310–321.

[13] Sheng, D.; Wei, Y.; Chen, Y.; Wang, Y. Convolutional neural networks with fractional order gradient method. Neurocomputing **2020**, 408, 42–50.

[14] Liu, J.; Chen, S.; Cai, S.; Xu, C. The Novel Adaptive Fractional Order Gradient Decent Algorithms Design via Robust Control. arXiv **2023**, arXiv:2303.04328.

[15] Liu, J.; Zhai, R.; Liu, Y.; Li, W.; Wang, B.; Huang, L. A quasi fractional order gradient descent method with adaptive stepsize and its application in system identification. Appl. Math. Comput. **2021**, 393, 125797.

[16] Hai, P.V.; Rosenfeld, J.A. The gradient descent method from the perspective of fractional calculus. Math. Methods Appl. Sci. **2021**, 44, 5520–5547.

[17] Haghighinia, A.; Tabatabaei, S.M.; Movahedirad, S. A novel geometrically-hybrid microchannel for performance enhancement in mass transfer: Description of Lyapunov exponent and Poincare map. Int. J. Heat Mass Transf. **2021**, 165, 120700.

[18] Narendra, K.S.; Annaswamy, A.M. Stable Adaptive Systems; Dover Publications Inc.: Mineola, NY, USA, **2005**.

[19] Kilbas, A.; Srivastava, H.; Trujillo, J. Theory and Applications of Fractional Differential Equations; Elsevier: Amsterdam, The Netherlands, 2006.

[20] Diethelem, K. The Analysis of Fractional Differential Equations; Springer: Berlin/Heidelberg, Germany, **2010**.

[21] Benavides, G.E.C.; Duarte-Mermoud, M.A.; Orchard, M.E.; Travieso-Torres, J.C. Pitch Angle Control of an Airplane Using Fractional Order Direct Model Reference Adaptive Controllers. Fractal Fract. **2023**, 7, 342.

[22] Narendra, K.S.; Annaswamy, A.M. Persistent excitation in adaptive systems. Int. J. Control **1987**, 45, 127–160.

[23] The MathWorks Inc. Control System Toolbox User's Guide. 1998. Available online: https://www.mathworks.com (accessed on 4 December 2024).

[24] Valerio, D.; da Costa, J.S. Ninteger: A non-integer control Toolbox for Matlab. In Fractional Derivatives and Applications; OIFAC: Bordeaux, France, **2004**.

[25] Sabatier, J.; Aoun, M.; Oustaloup, A.; Gregoire, G.; Ragot, F.; Roy, P. Fractional system identification for lead acid battery state of charge estimation. Signal Process. **2006**, 86, 2645–2657.

[26] Clerc, M. Particle Swarm Optimization; ISTE Ltd.: London, UK, **2006**.

www.ingramcontent.com/pod-product-compliance
Lightning Source LLC
LaVergne TN
LVHW051131210926
848482LV00001B/64